Worth the Wait

DISCOVER HOW GOOD THINGS TAKE TIME IN THIS ANTHOLOGY OF SCIENCE STORIES:

IN 1 MINUTE – A human heart beats

IN 8 MINUTES – Light from the sun reaches Earth

IN 1 HOUR - A barn owl hunts its prey

IN 1 NIGHT – The moon rises and falls

IN LESS THAN A DAY - A dragonfly grows its wings

IN 10 DAYS – Starlings incubate their eggs

IN TWO WEEKS - A camel crosses the desert

IN 1 MONTH – A lunar cycle occurs

OVER MANY MONTHS - A honeypot ant stores its food

IN 1 WINTER – A bear waits for spring

IN 9 MONTHS – A baby is born

IN 1 YEAR – An apple tree grows fruit

IN MORE THAN 25 YEARS – A coral reef grows

IN 30 YEARS - A century plant blooms

IN MORE THAN 60 YEARS - A patch of rainforest regrows

IN 70 YEARS - An elephant lives a lifetime

IN 80 YEARS – A human is born, lives and dies

IN 100 YEARS - A giant tortoise completes its lifecycle

MAGIC CAT ✸ PUBLISHING

IN ONE WINTER

Bear waits for spring

In the depths of an alpine forest, through the great pine and spruce trees, there is a den, barely visible for the rocks and fallen branches that secure its entrance. Deep within this cosy home, Grizzly Bear is settling in for the winter.

He has been busy all year, building up his fat reserves and digging and fetching materials from the forest for bedding. Now he retreats into a type of slumber, hibernating for between four and seven months without the need for food or water.

Just as we wait for the ground to thaw and show promise of life, Grizzly Bear is fast asleep until spring arrives again...

Autumn ends...

Spring is here.

3

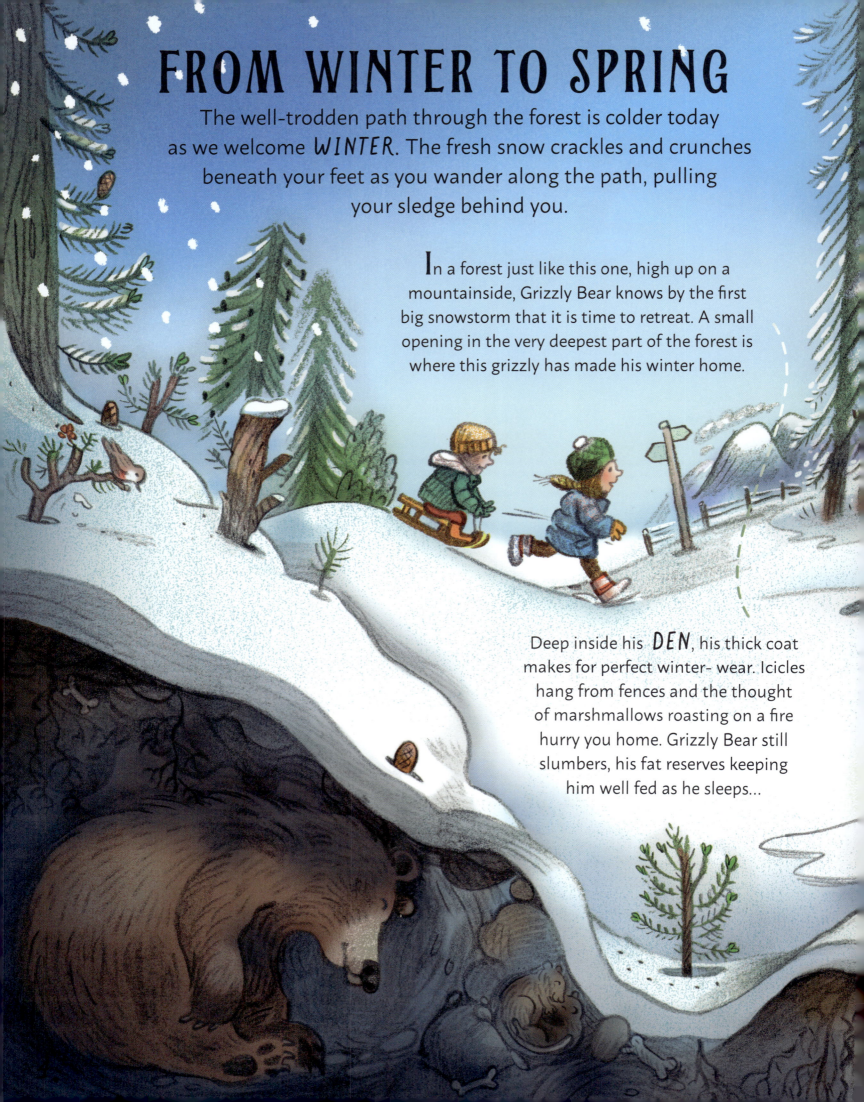

FROM WINTER TO SPRING

The well-trodden path through the forest is colder today as we welcome **WINTER**. The fresh snow crackles and crunches beneath your feet as you wander along the path, pulling your sledge behind you.

In a forest just like this one, high up on a mountainside, Grizzly Bear knows by the first big snowstorm that it is time to retreat. A small opening in the very deepest part of the forest is where this grizzly has made his winter home.

Deep inside his **DEN**, his thick coat makes for perfect winter- wear. Icicles hang from fences and the thought of marshmallows roasting on a fire hurry you home. Grizzly Bear still slumbers, his fat reserves keeping him well fed as he sleeps...

As the months pass by, snowy pathways melt away. The den remains silent and still... but in the valley, the birds start to busy themselves.

From spring to autumn, hungry **GRIZZLY BEARS** use the warmer weather to gather food sources and build up their fat reserves, which will enable them to survive the winter.

Vivid green grass and wildflowers now line the path. Finally, high up in the still-snowy mountains, Grizzly Bear is waking up...

As you turn to face the *SPRING* sunshine, Grizzly Bear merges from his den. He yawns as he breathes in the warming air, before venturing out onto the thawing snow, one slow, sleepy step at a time.

5

IN ONE YEAR

An apple tree grows fruit

An apple tree is growing and changing all the time —
just like you and me. It takes a whole year for an apple
to transform from bud to fruit, over four busy seasons.

Every year, the apple tree will produce a different
amount of fruit. To thrive, it will need plenty of sun,
a little shade and well drained, moist soil.

It cleverly knows exactly what to do each season,
to work with the weather and start its life cycle again,
and again.

7

Laura Brand

Laura Brand is an award-winning author and creator of *The Joy Journal*, which has hundreds of thousands of followers all over the world. A nature educator and avid mindfulness practitioner, this is her second book for children.
@thejoyjournal

Leonie Lord

After studying illustration at Central St. Martin's in London, Leonie Lord began her children's picture book career. She has illustrated over 15 picture books, and lives with her husband and two sons in Gloucestershire.